5th Grade Math
Volume 6

© 2013 OnBoard Academics, Inc
Newburyport, MA 01950
800-596-3175
www.onboardacademics.com

ISBN: 978-1494857257

Table of Contents

Exploring Symmetry   4

Exploring Symmetry Quiz   10

Angle Properties   11

Angle Properties Quiz   15

Units of Measure   16

Units of Measure Quiz   26

Area   27

Area Quiz   31

Perimeter and Area of Irregular Figures   32

Perimeter and Area of Irregular Figures Quiz   38

Exploring Symmetry

Key Vocabulary

line symmetry

rotational symmetry

congruent/congruence

Draw the lines of symmetry for this rectangle.

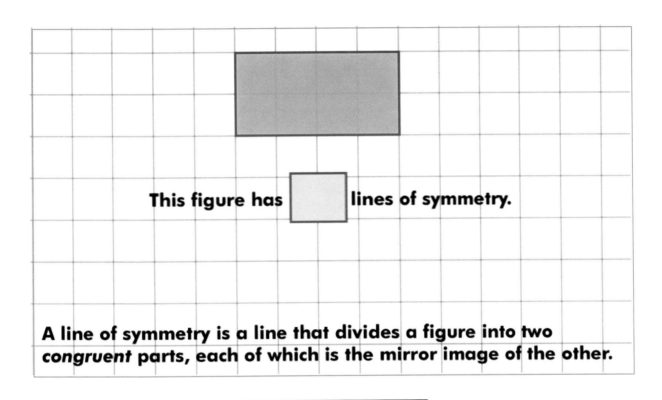

This figure has [ ] lines of symmetry.

A line of symmetry is a line that divides a figure into two congruent parts, each of which is the mirror image of the other.

Study the triangles' lines of symmetry and label the triangle types.

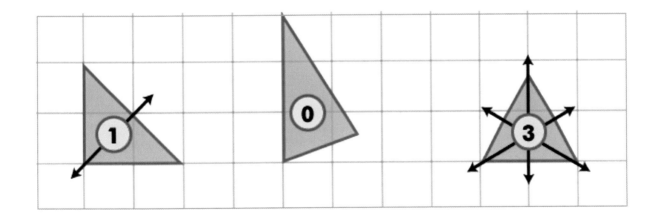

Jesús was asked to draw lines of symmetry for these quadrilaterals. Circle the ones drawn correctly.

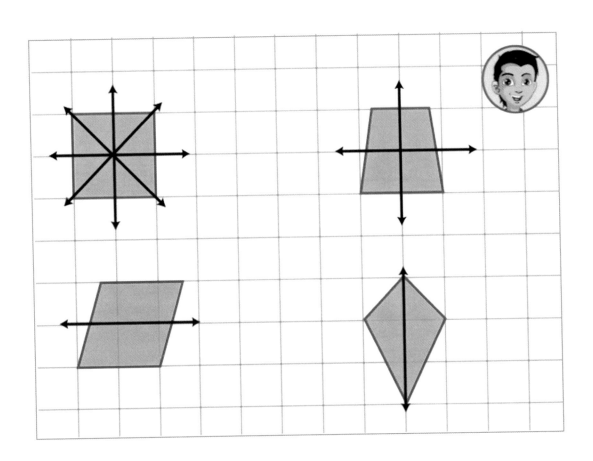

Rotational Symmetry

A figure has rotational symmetry if it looks the same at any point when it is rotated less than a full turn.

The top of the triangle is noted by a red dot. Notice the shape as it rotates counter clockwise. Circle the positions of rotational symmetry.

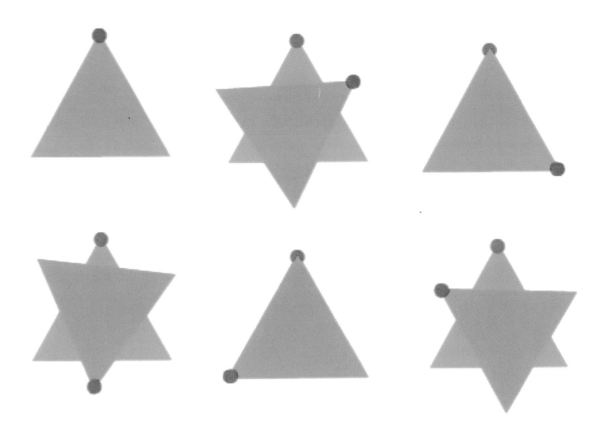

The equilateral triangle has rotational symmetry at these angles.

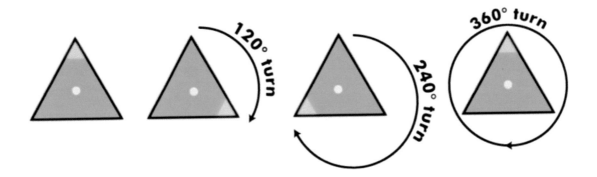

120° and 240°. At 360° the triangle has returned to its original position.

Circle the figures that have rotational symmetry.

Complete these shapes so that they have rotational symmetry.

Name: _____

Exploring Symmetry Quiz

1 True or false? This figure has rotational symmetry.

2 Which shape has no lines of symmetry?

3 How many lines of symmetry does this shape have?

4 How many lines of symmetry does this shape have?

Angle Properties

Key Vocabulary

right angle

obtuse angle

acute angle

supplementary angles

protractor

Draw a line to connect the angles with its correct angle facts.

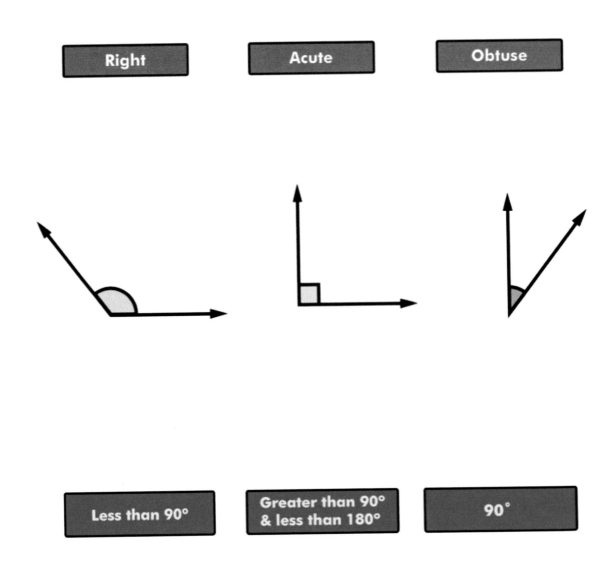

Angle Estimation Game

This game can be played in teams or individually. If completing individually enter your answers into the Team 1 line.

OVERVIEW
1. Divide the class into two teams: Team 1 and Team 2.
2. Ask each Team to estimate the measure of each of the 3 angles.
3. Record the estimates in the left side of the table cells. You can record the score that the Team receives for each estimate in the right side of the table cells (see Scoring below).

SCORING

Award 5 points for an estimate that is within 5˚ of the correct answer.

Award 2 points for an estimate that is within 10˚ of the correct answer.

Award 0 points for an estimate that is incorrect by more than 10˚.

Deduct 2 points for an estimate that is incorrect by more than 20˚.

Sum the points to determine the winning Team.

MEASURING THE ANGLES

When the Teams have made their estimates, copy and paste the protractor from this page to measure the angles. Draw a rectangle using the Magic Pen Tool to zoom in on the measurement.

Note that the angle images are locked and so students will not be able to rotate them to make their estimates, but they can rotate the protractor to measure the angles.

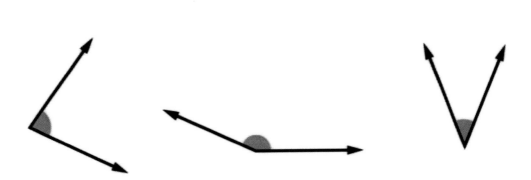

	Angle 1	Angle 2	Angle 3	Total
Team ①	°	°	°	
Team ②	°	°	°	

The sum of the measure of the angles on a straight line is 180⁰

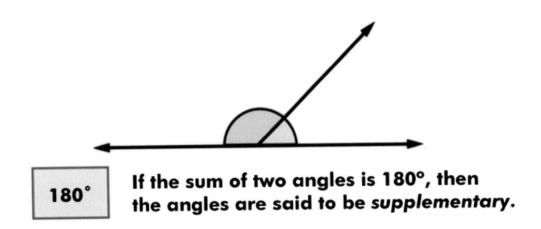

180° | If the sum of two angles is 180°, then the angles are said to be *supplementary*.

Match the angles and the segments.

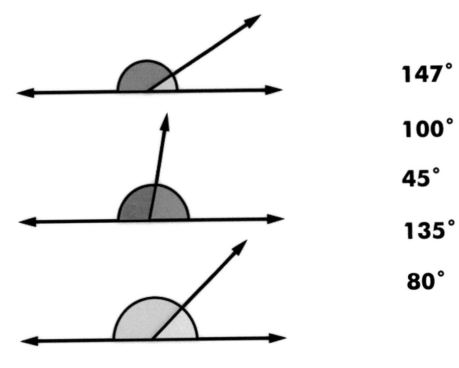

147°

100°

45°

135°

80°

What is the measure of the missing angle?

Name: _____

Angle Properties Quiz

① **Are these angles correctly classified?**

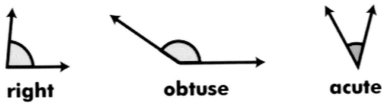

right   obtuse   acute

② **What is the measure of angle n?**

Ⓐ **16°**

Ⓑ **196°**

Ⓒ **14°**

Ⓓ **106°**

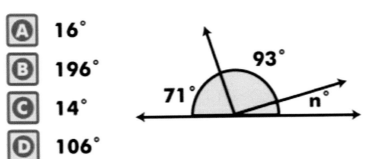

71°   93°   n°

③ **What is the measure of angle a (in degrees)?**

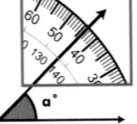

a°

Units of Measure

Key Vocabulary

metric measure

customary measure

length

mass

weight

Can you order these customary and metric units of measure?

Customary Units	
Weight	**Length**

GREATEST

LEAST

Metric Units	
Mass	**Length**

Millimeter　Pound　　Kilogram　Meter　Ton　Centimeter

Kilometer　Foot　Yard　　Ton　　Ounce　Gram　Mile　Inch

Draw a line to connect the equal customary units of measure.

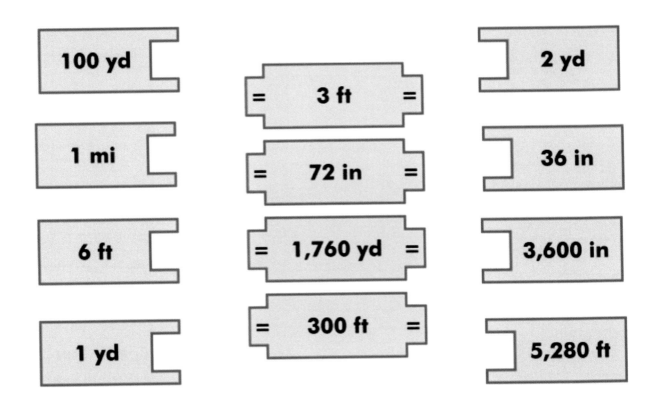

Measuring Lengths: Customary units of measure.

User your ruler to measure these items.

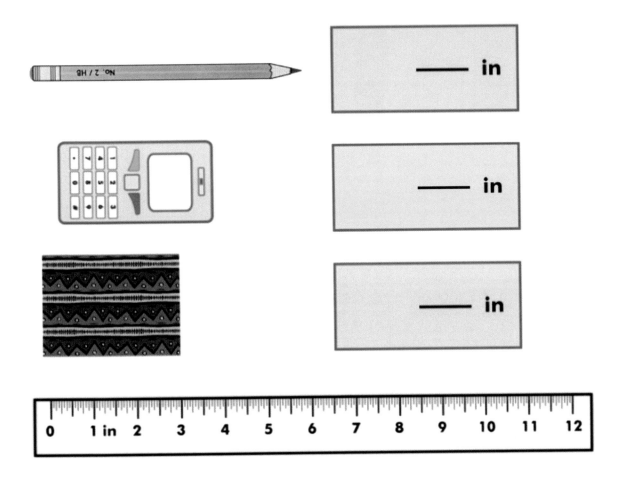

Can you match each image with an appropriate metric measurement.
Write you answer in the blue box.

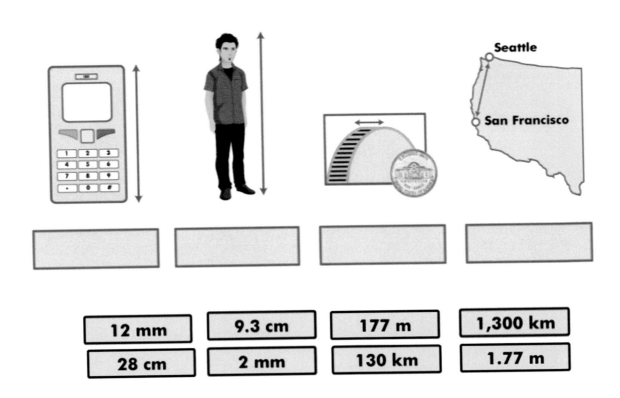

| 12 mm | 9.3 cm | 177 m | 1,300 km |
| 28 cm | 2 mm | 130 km | 1.77 m |

Measure the lengths of these objects to the nearest Millimeter.

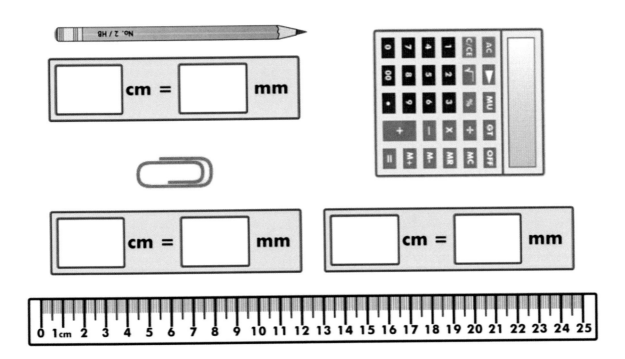

Customary units of weight.

Match the item with its approximate weight.

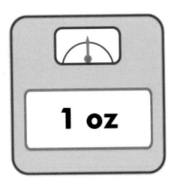
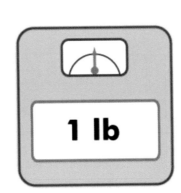
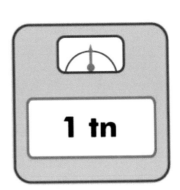

What is the most appropriate unit of measure for the mass of each object below?

56 kg   0.4 t   20 g   2 g   5.6 kg   4 t

Approximate customary to metric measure conversions

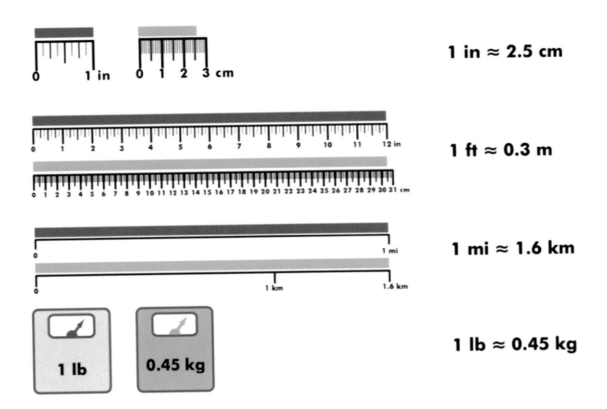

1 in ≈ 2.5 cm

1 ft ≈ 0.3 m

1 mi ≈ 1.6 km

1 lb ≈ 0.45 kg

Practice converting between customary and metric measurement units.

5 in ≈ [ ] cm?

3 ft ≈ [ ] m?

8 mi ≈ [ ] km?

$7\frac{1}{2}$ lb ≈ [ ] kg?

Name: _____

Units of Measure Quiz

1 True or false, a mile is shorter than a kilometer?

2 Which of the following statements is *not* correct?

- Ⓐ 10 ft = 120 in
- Ⓑ 17,600 ft = 10 mi
- Ⓒ 72 in = 2 yd
- Ⓓ 5 yd = 15 ft

3 5.5 m = [?] cm?

4 12 in ≈ [?] cm?

30	20	10	48
Ⓐ	Ⓑ	Ⓒ	Ⓓ

Area

Key Vocabulary

area

Parallelogram

square unit

The area of this rectangle is 8 square inches.

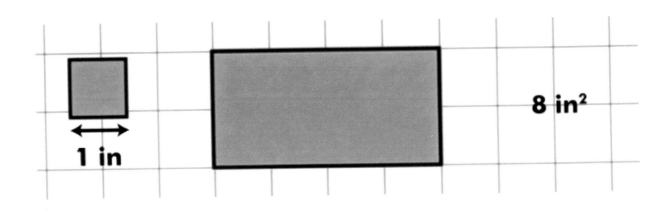

To measure the parallelogram how would you make a rectangle?

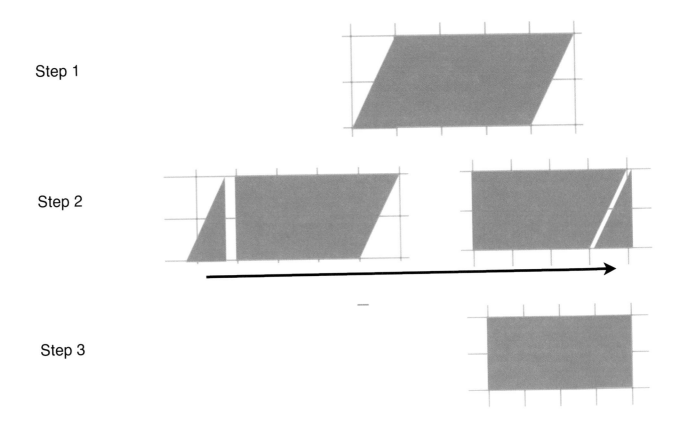

Step 1

Step 2

Step 3

The formula for the area of a parallelogram.

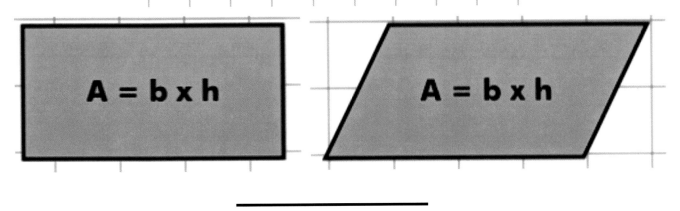

Rectangles and parallelograms with the same base and the same height have the same area.

A = b x h

A = b x h

Find the area for the yellow triangles.
The dotted lines provide a hint.

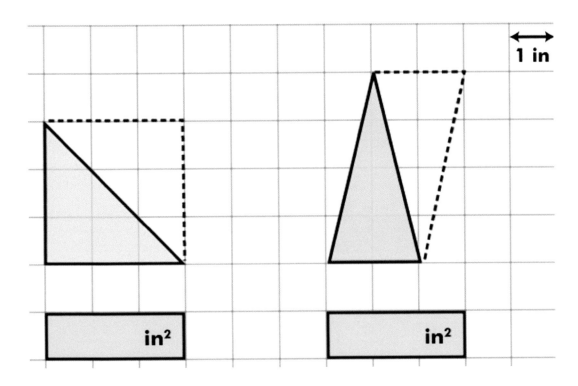

1 in

in²

in²

The area of a triangle is half the area of a parallelogram.

Triangles and half parallelograms
Study the illustration below to discover the the formula for the area of a triangle.

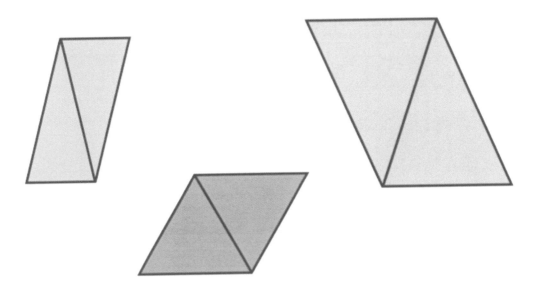

Area of a parallelogram = base x height

Area of a triangle = $\frac{1}{2}$ (base x height)

Name: _____

Area Quiz

1 True or false? The area of a parallelogram is half the area of a triangle.

2 What is the area of Figure 1?

     (A) 7.2 cm^2 **(B)** 4.8 cm^2 **(C)** 7.8 cm^2 **(D)** 6.2 cm^2

3 What is the area of Figure 2 in sq ft?

4 What is the length of the base of Figure 3 in feet?

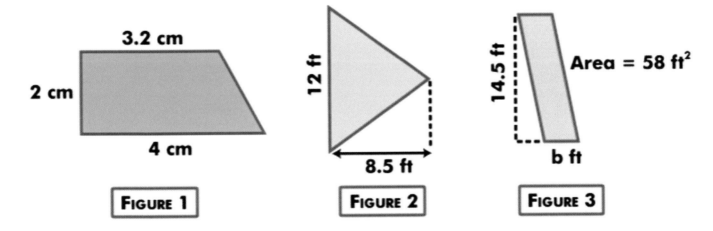

FIGURE 1 — 3.2 cm (top), 2 cm (left side), 4 cm (bottom)

FIGURE 2 — 12 ft (height), 8.5 ft (base)

FIGURE 3 — 14.5 ft (height), b ft (base), Area = 58 ft²

Perimeter and Area of Irregular Figures

Key Vocabulary

perimeter

area

square units

Find the perimeter of this room.

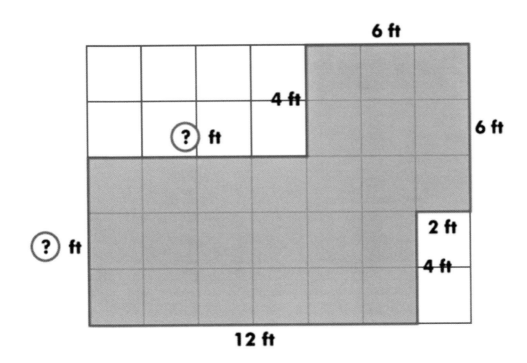

Find the area of this room.

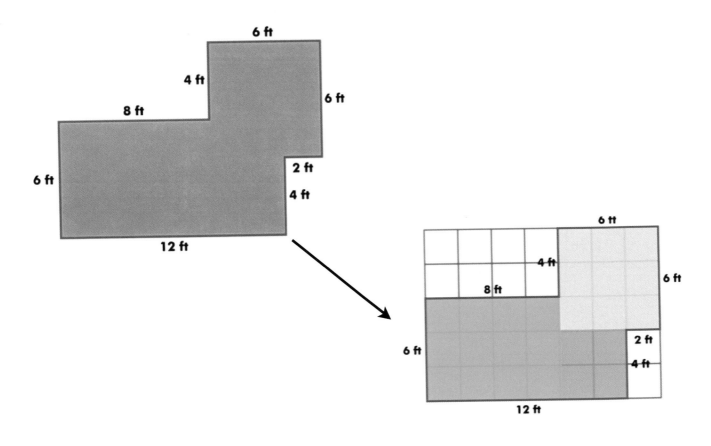

Area of green rectangle [　　　] ft²   **Total area of room** [　　　] ft²

Area of blue rectangle [　　　] ft²

Area of yellow rectangle [　　　] ft²

Find the perimeter and area for these shapes.

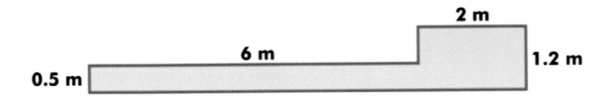

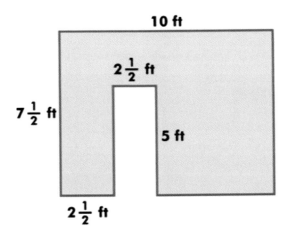

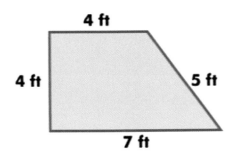

Find the perimeter and area of Mrs. Jones' apartment.

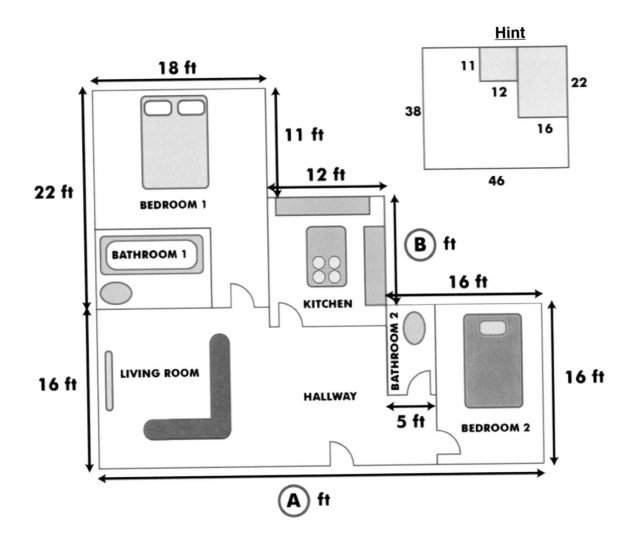

Hint

Estimate the perimeter and area of this shape.

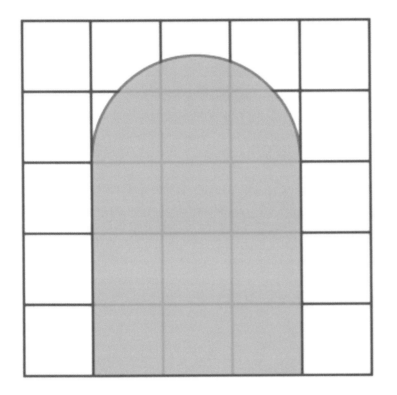

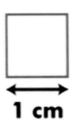

1 cm

Name: _____

Perimeter and Area of Irregular Figures Quiz

(1) **The perimeter of this rectangle is 2a + 2b. True or false?**

a [b]

(2) **Which are all units of area?**

(A) ft^2, m^2, square yard, square mile, acre

(B) feet, centimeter, meter, mile, yard, inch

(C) ft^2, m^2, cm^3, mm^3

(D) ft^3, m^3, cm^3, mm^3

(3) **The perimeter of this figure is 6a + ___ b**

2b
3a

(4) **The area of this figure is ___ ab**

Newburyport, MA 01950

1-800-596-3175

OnBoard Academics employs teachers to make lessons for teachers! We create and publish a wide range of aligned lessons in math, science and ELA for use on most EdTech devices including whiteboard, tablets, computers and pdfs for printing.

All of our lessons are aligned to the common core, the Next Generation Science Standards and all state standards.

If you like our products please visit our website for information on individual lessons, teachers licenses, building licenses, district licenses and subscriptions.

Thank you for using OnBoard Academic products.

Made in the USA
Las Vegas, NV
28 January 2022

42540280R00026